ARBORETUM ET F
BRITANNICUM;

OR,

THE TREES AND SHRUBS OF BRITAIN,

Native and foreign, hardy and half-hardy,

PICTORIALLY AND BOTANICALLY DELINEATED,
AND SCIENTIFICALLY AND POPULARLY DESCRIBED;

WITH

THEIR PROPAGATION, CULTURE, MANAGEMENT,
AND USES IN THE ARTS, IN USEFUL AND ORNAMENTAL PLANTATIONS, AND IN
LANDSCAPE-GARDENING;

PRECEDED BY A
HISTORICAL AND GEOGRAPHICAL OUTLINE
OF THE TREES AND SHRUBS OF TEMPERATE CLIMATES
THROUGHOUT THE WORLD.

By J. C. LOUDON, F.L. & H.S., &c.

AUTHOR OF THE ENCYCLOPÆDIAS OF GARDENING AND OF AGRICULTURE,
AND CONDUCTOR OF THE GARDENER'S MAGAZINE.

IN EIGHT VOLUMES:
FOUR OF LETTERPRESS, ILLUSTRATED BY ABOVE 2500 ENGRAVINGS;
AND FOUR OF OCTAVO AND QUARTO PLATES.

VOL. V.
THE PLATES FROM MAGNOLIA'CEÆ TO LEGUMINO'SÆ INCLUSIVE.

LONDON:
PRINTED FOR THE AUTHOR;
AND SOLD BY
LONGMAN, ORME, BROWN, GREEN, AND LONGMANS;
THE PARTIALLY COLOURED AND COLOURED COPIES, BY
JAMES RIDGWAY AND SONS.
1838.

Àcer créticum.

The Cretan Maple.

Full grown tree at Syon, 29 ft. high; diam. of the trunk, 2½ ft. and of the head, 48 ft.
[Scale 1 in. to 12 ft.]

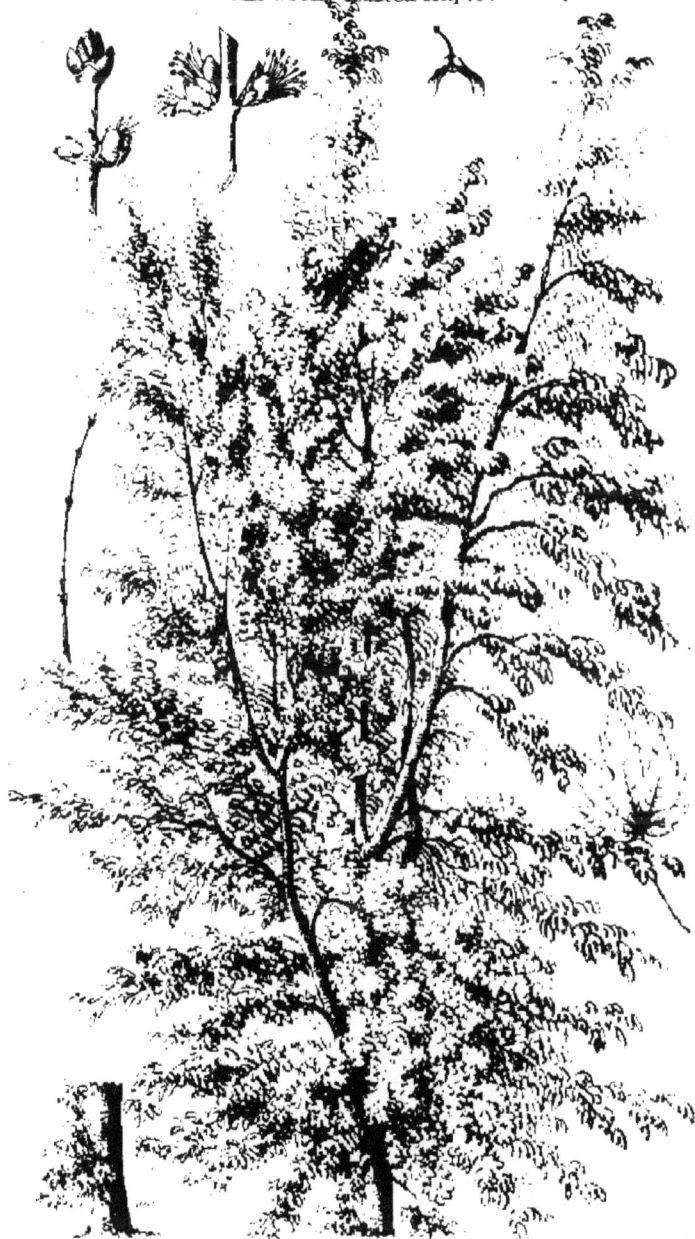

Àcer eriocárpon.

The woolly-fruited Maple.

Full-grown tree at Kew, 50 ft. high; diam. of the trunk, 3 ft.; and of the head, 15 ft.
[Scale 1 in. to 12 ft.]

Acer macrophyllum.

Clerum broad-stipl

Acer monspessulànum .

The Montpelier Maple .

Àcer monspessulànum.

The Montpelier Maple.

Full grown tree at Ham House, 30 ft. high ; diam. of the head, 40 ft. ; and of the trunk, 1½ ft.
[Scale 1 in. to 12 ft.]

Acer *hýbridum*.
The hybrid Maple.

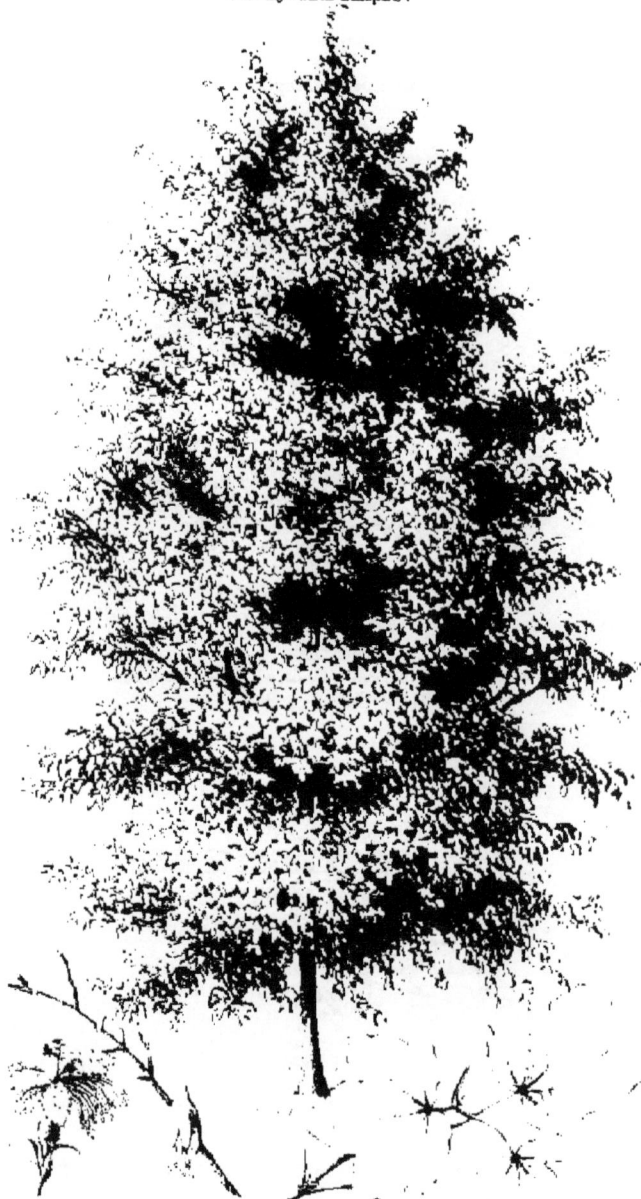

`Acer Ópulus.
The Guelder-rose-*like* Maple.

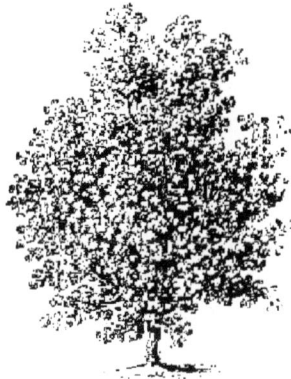

10 ft. high. 3½ in. diam.

IV. r.

Acer opulifolium.

The Guelder-Rose-leaved Maple.

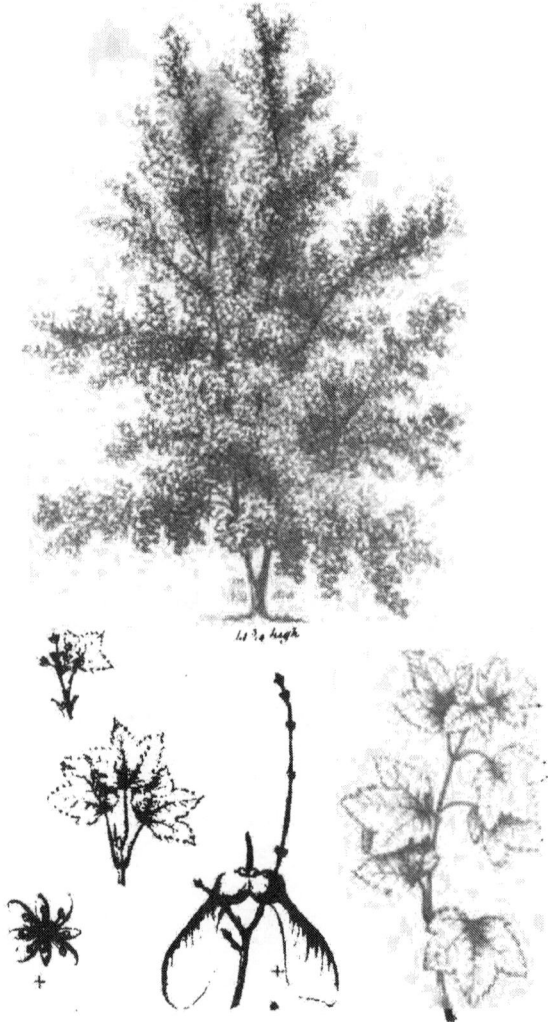

44 ½ a high

Printed from Zinc by Day & Haghe

Acer platanöides.
The Platanus-like Maple.

A`cer platanoïdes laciniátum..
The cut-*leaved* Platanus-like Maple.

Printed from Zinc by Day & Haghe.

IV.H
A'cer Pseudo-Plátanus.
The *Sycamore*, or Bastard Plane tree Maple.

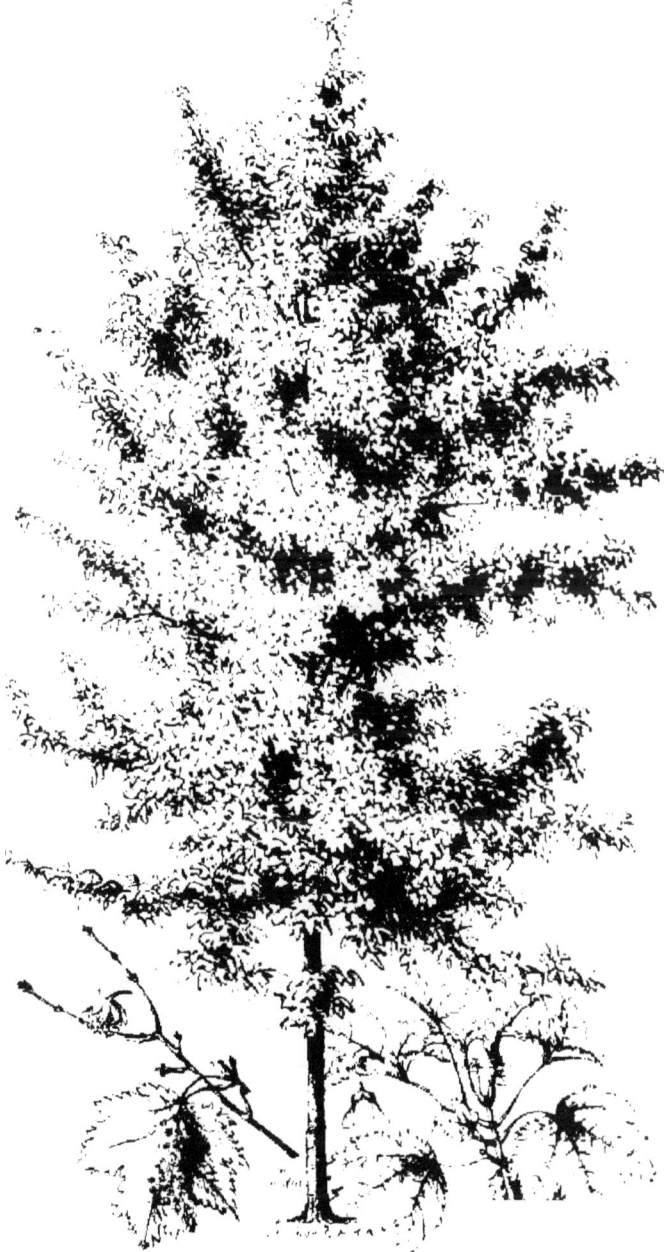

Àcer Pseùdo-Plátunus.

The false Platanus, *or common* Sycamore.

Full-grown tree at Studley, 100 ft. high; diam. of the trunk, 8 ft. 1 in.; and of the head, 91 ft

[Scale 1 in. to 24 ft.]

IV. o.
A'cer rúbrum.
The red-*flowered* Maple.

Printed from Zinc by Day & Haghe.

Àcer rùbrum.

The Red, *or Swamp* Maple.

Full grown tree at Kew, 26 ft. high; diam. of the trunk, 2½ ft ; and of the head, 30 ft.
[Scale 1 in. to 12 ft.]

Printed from Zinc by Day & Haghe

IV.E.
Acer spicatum.
The spike-*flowered* Maple.

Printed from a Zinc by Day & Haghe

Acer striatum.
The striped-*barked* Maple.

A`cer tatáricum.
The Tartarian Maple.

Printed from /sinch, Day & Hague.

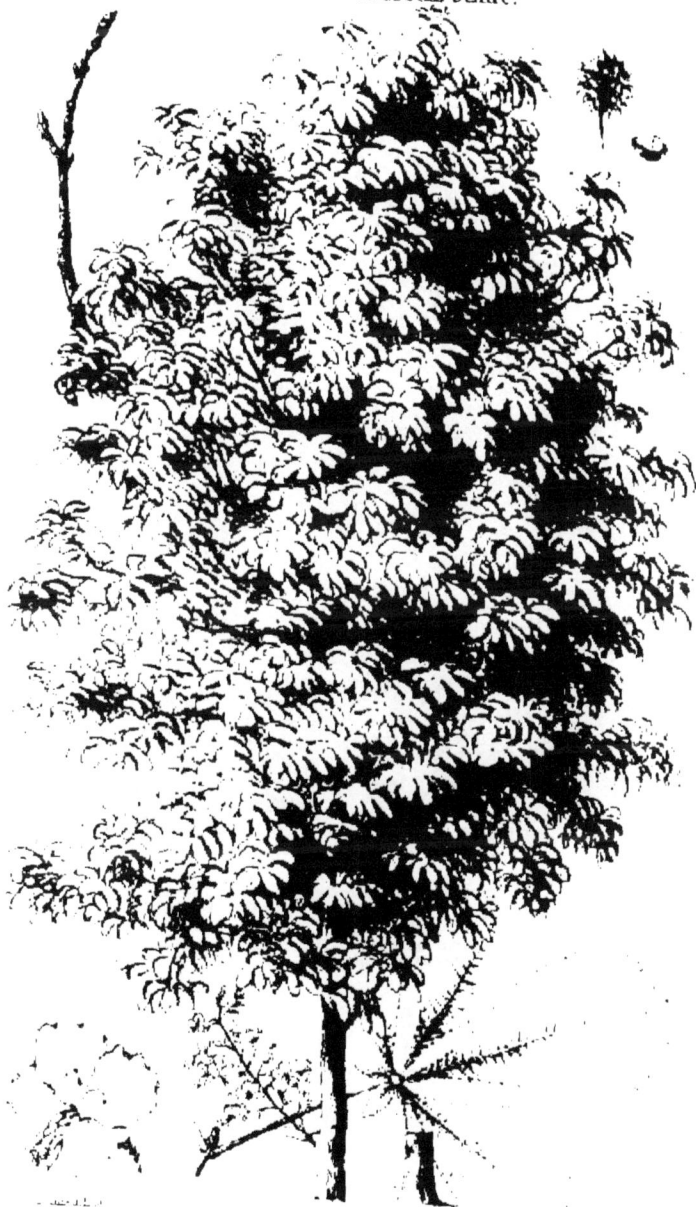

VI.A
Æsculus Hippocástanum.
The common Horsechestnut.

43

Æsculus Hippocastanum.
The *common* Horse-Chestnut.

Full-grown tree at Forty Hill, Enfield, 62 ft. high; diam. of the trunk, 4 ft.; and of the head, 48 ft.
[Scale 1 in. to 12 ft.]

Printed from J.D. by ... y & Hughs

Ailántus *glandulòsa.*

The glandulous-*leaved* Ailantus.

Ailántus *glandulòsa*.

The glandulous-*leaved* Ailantus.

Full-grown tree at Syon, 71 ft. high.
[Scale 1 in. to 12 ft.]

Aristotèlia Mácqui.

The Macqui Aristotelia.

10 ft. high, 1½ in. diam.

Caragána *arboréscens.*
The arborescent Siberian Pea tree.

Printed there June by Day & Haghe

XXV.A.
Cercis Siliquastrum.
The common Judas tree.

ft. high.

Printed from Zinc. by Day & Haghe

Cércis Siliquástrum.

The *common* Judas tree.

Full grown tree at Syon, 24 ft. high ; diam. of the trunk, 2 ft. ; and of the head, 22 ft.
[Scale 1 in. to 12 ft.]

Cytisus alpinus.
The alpine Cytisus, *or Scotch Laburnum.*

Cýtisus L. alpinus péndulum.
The pendulous Alpine, *or Scotch*, Laburnum.

12 ft. high, 2½ in. diam.

Cytisus Laburnum.
The *common* Laburnum.

Cytisus Laburnum incisum.
The cut-*leaved* Laburnum .

Printed from Zinc : by Day & Haghe

Euónymus európaeus.
The European, *or common* Spindle tree.

12 high

Gleditschia (triacánthos) inérmis.

The unarmed *or thornless* Gleditschia, *or Honey Locust.*

Gleditschia inérmis.

The unarmed Gleditschia, *or Honey Locust Tree.*

Full grown tree at Syon, 72 ft. high; trunk, 2 ft. 4 in. diam.; head, 71 ft diam
[Scale 1 in. to 12 ft.]

Gleditschia macracántha.
The long-spined Gleditschia, *or Honey Locust.*

Full-grown tree at S) on, 57 ft. high ; diam. of the trunk, 3 ft. ; and of the head, 63 ft.
[Scale 1 in. to 12 ft.]

Gledítschia japónica.
The Japan Gleditschia.

10 ft. high, 2½ in. diam.

XXIII.

Gledítschia hórrida.

The horrid-*spined* Gleditschia, *or Honey Locust.*

Full-sized tree from Syon; 54 ft. high; trunk, 3 ft. diam.; diam. of the head, 54 ft.

Gledítschia hórrida purpúrea.

The purple horridly-*spined* Gleditschia, *or Honey Locust.*

12 ft. high, 2 in. diam.

XXIII.A.
Gleditschia triacánthos.
The three-spined *Honey Locust* or Gleditschia

Gleditschia triacánthos.

The three-thorned Gleditschia, *or Honey Locust.*

Full-grown tree at Syon, 68 ft. high; diam. of the trunk, 2 ft. 3 in.; and of the head, 40 ft.
[Scale 1 in. to 12 ft.]

Gymnócladus canadénsis.
The Canadian Gymnocladus, *or Kentucky Coffee tree.*

Gymnócladus canadénsis.

The Canadian Gymnocladus, *or Kentucky Coffee tree.*

Full-grown tree at Syon; 57 ft. high; diam. of the trunk, 8 ft.; and of the head, 47 ft.
[Scale 1 in. to 12 ft.]

Ìlex Aquifòlium.
The sharp-leaved, *or common,* Holly.

14 ft. high, 8½ in. diam.

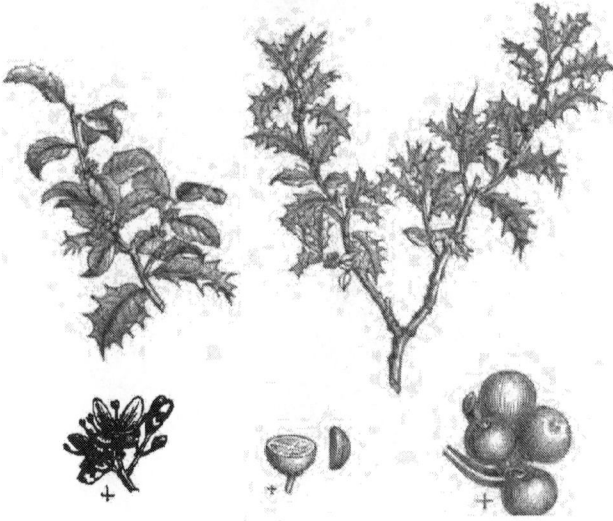

Ìlex Aquifòlium.
The sharp-leaved, *or common,* Holly.

Indigenous tree; as usually about 7 m. high.

[Scale 1 in. to 12 ft.]

Ilex opàca.

The opaque-*leaved* Holly.

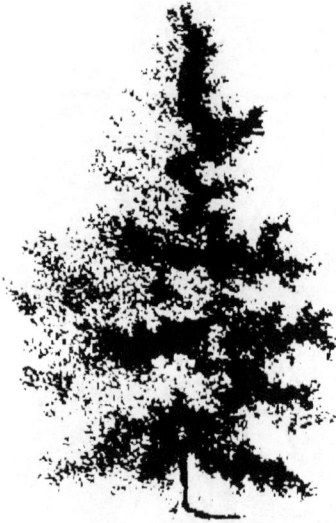

Ílex opàca.

The opaque-*leaved* Holly.

Full-grown tree at Syon, 18 ft. high ; diam. of the trunk, 1 ft. ; and of the branches, 23 ft.
[Scale 1 in. to 12 ft.]

Kölreutéria paniculàta.
The panicled-*flowering* Kölreuteria.

Printed from Nature by Day & Haghe

Liriodéndron Tulipífera.
The Tulip tree.

Printed in colour by Day & Haghe.

Liriodéndron Tulipífera.

The tulip-bearing Liriodendron, *or the Tulip tree.*

Full-grown tree at Syon, 76 ft. high.
[Scale 1 in. to 12 ft.]

Magnólia acumináta.

The pointed-*leaved deciduous* Magnolia.

Printed for no Time by Day & Hughes.

Magnòlia acuminàta.

The pointed-*leaved* Magnolia.

Full-grown tree at Syon; 49 ft. high; diam. of the trunk, 3 ft.; and of the head, 49 ft.
[Scale 1 in. to 12 ft.]

1.H.

Magnòlia auriculàta.

The ear-leaved *deciduous* Magnolia.

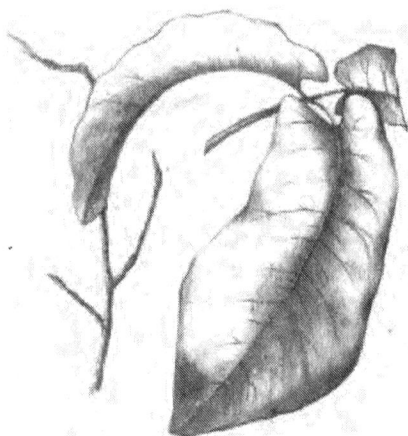

Printed from Dawson's Lithography

Magnòlia glaúca.
The glaucous-*leaved deciduous* Magnolia.

Printed from Zinc by Day & Haghe.

Magnolia glauca Thompsoniana.

Thompson's glaucous-*leaved deciduous* Magnolia.

I.M.

Magnòlia conspícua.

The conspicuous-*flowered deciduous* Magnolia.

Magnòlia cordàta.
The heart-shaped-*leaved deciduous* Magnolia.

Printed from Zinc by Day & Haghe

Magnòlia grandiflòra.
The large-flowered *evergreen* Magnolia.

Drawn nom. nat. by J. W. Kingda

Magnolia grandiflora ferruginea.
The large-flowered evergreen rusty-leaved Magnolia.

Printed from Zinc by Day & Haghe.

Magnòlia grandiflora exoniénsis.
The large-flowered *evergreen* Exeter Magnolia.

I.G.

Magnolia macrophylla.
The large-leaved *deciduous* Magnolia.

Printed from Nat. by Day & Haghe.

Magnólia pyramidàta.
The pyramidal *deciduous* Magnolia.

Magnòlia tripétala.
The three-petaled-*flowered deciduous* Magnolia.

V.A.
Negúndo *fraxinifolium.*
The Ash-leaved Box Elder.

V.B.

Negúndo fraxinifólium crispum.

The curled Ash-leaved Box Elder.

Printed from Zinc by Day & Haghe.

Paliurus aculeatus.
The prickly Christ's Thorn.

Printed from Zinc by Day & Haghe.

Paliùrus aculeàtus.

The prickly Paliurus, *or Christ's thorn.*

Full-grown tree at Syon; 33 ft. high; diam. of the trunk, 1 ft. and of the head, 30 ft.
[Scale 1 in. to 12 ft.]

Pàvia flàva.

The yellow-*flowered*, Pavia, *or smooth-fruited Horse-Chestnut.*

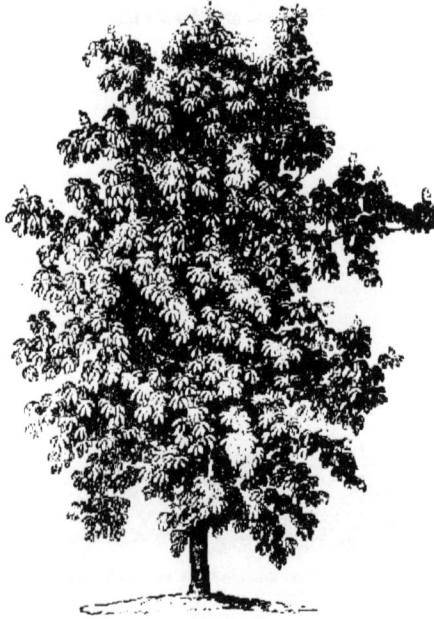

16 ft. high, 6 in. diam.

Æ'sculus (*Pàvia*) *flàva.*
The yellow-*flowered* (Pavia) Horse-chestnut.

Full-grown tree at Syon; 40 ft. high; trunk, 1 ft. 4 in. diam.; diam. of the head, 20 ft.

Pàvia macrocárpa.
The large-fruited Pavia.

13 ft. high, 3 in. diam.

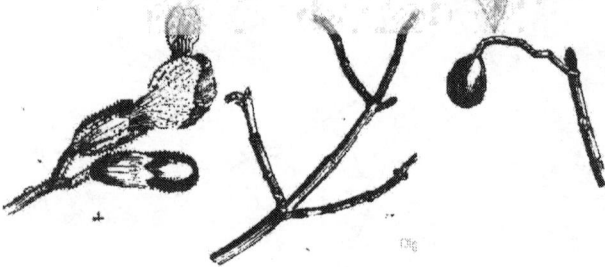

Pàvia rùbra.

The red-*flowered* Pavia, or *smooth-fruited, Horse-chestnut tree.*

10 ft. high. 5½ in. diam.

Pàvia rùbra.

The red-*flowered* Pavia, *or smooth fruited, Horse-chestnut tree.*

Full grown tree at Syon ; 26 ft. high ; several small trunks from the same root head, 27 ft. diam.
[Scale 1 in. to 12 ft.]

Pàvia húmilis péndula.
The pendulous-*branched* Pavia.

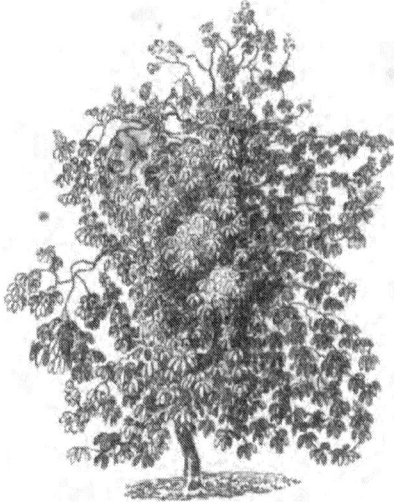

11 ft. high, 2¾ in. diam.

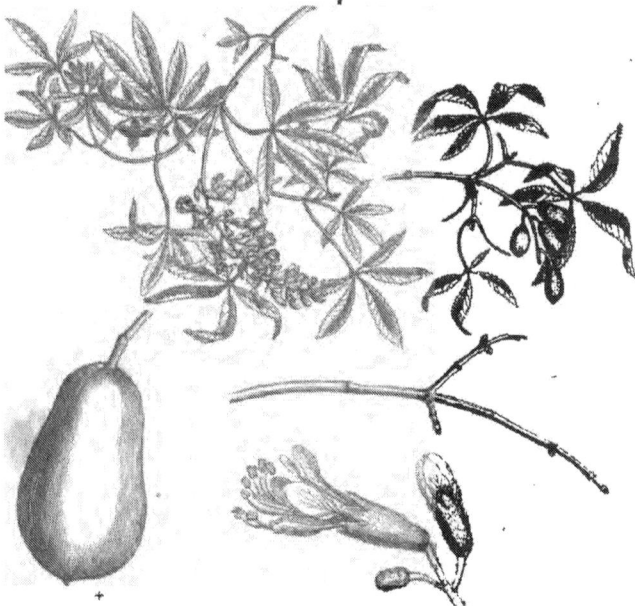

Rhámnus álpinus.
The Alpine Rhamnus, *or Buckthorn.*

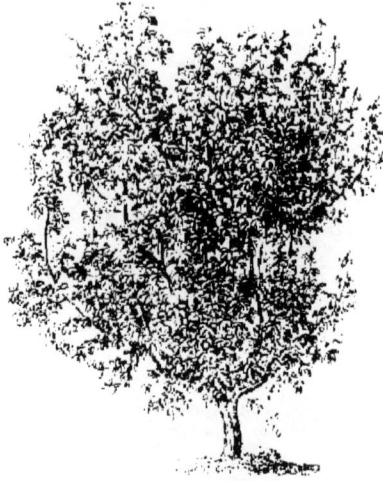

11 ft. high, 3 tn. diam.

Rhámnus catháríicus.
The purging Buckthorn.

Printed from Zinc by Day & Haghe.

Rhámnus Frángula.

The Frangula, *or brittle-wooded*, Buckthorn.

30 ft. high, 1½ in. diam.

Rhámnus latifólius.

The broad-leaved Buck-thorn.

11 ft. high, 3 ½ in. diam.

Robínia (híspida) macrophýlla.
The long-leaved Robinia, *or Rose Acacia.*

8½ ft. high, 2½ in. diam.

XX.A.
Robinia Pseud-Acàcia.
The *common* Locust, *or* Bastard Acacia.

Robínia Pseùd-Acàcia tortuòsa.

The twisted-leaved Robinia or False Acacia.

12 ft. high, 3½ in. diam.

Robinia Pseud-acàcia umbraculífera.
The Umbrella, or *Parasol*, False Acacia Robinia.

XX. c
Robinia viscòsa.
The viscid-*barked* Locust.

Printed from Zinch; Day & Haghe.

XVII. A.

Sophòra japónica.
The Japanese Sophora.

20 ft. high, 9 in. diam.

Sophòra *japónica péndula*.
The drooping-*branched* Japanese Sophora.

13 ft. high, 3 in. diam.

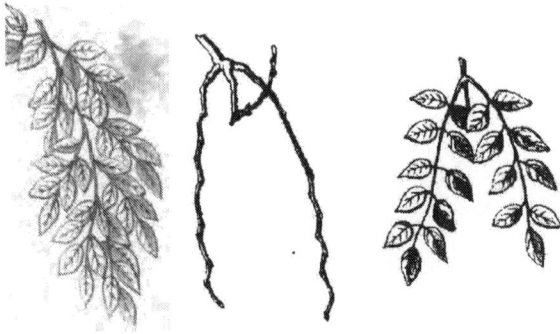

Tília álba.

The white-*leaved* Lime tree.

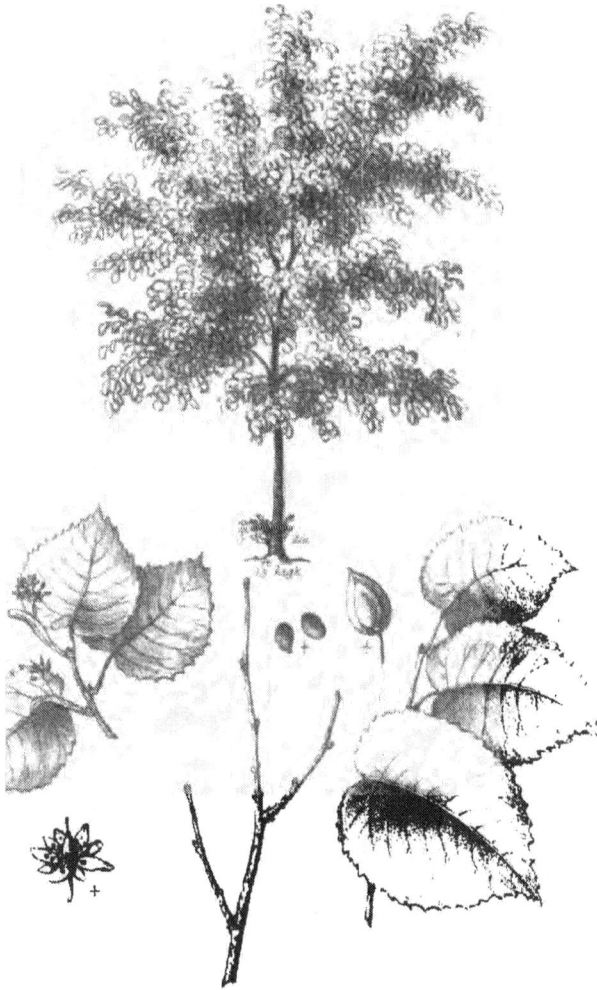

Tilia argéntea.
The silvery-*leaved* Lime tree.

Printed from Zinc by Day & Haghe.

Printed from Zinc by Day & Haghe.

III.A.

Tilia europæa.

The European, *or common,* Lime Tree.

Tilia europæa.

The European, *or common*, Lime tree.

Full-grown tree at Studley Park, 128 ft. high; diam. of the trunk, 6 ft. ; and of the head, 75 ft.
[Scale 1 in. to 24 ft.]

Printed from Zinc by Day & Haghe.

Tilia platyphylla.

The broad-leaved Lime tree.

Tília europæa platyphýlla mìnor.

The smaller broad-leaved European Lime tree.

15 ft. high, 4 in. diam.

XVIII.A.

Virgilia lutea.

The yellow-*flowered* Virgilia.

13 high.

Printed & Published by I. & N. d...

Xanthóxylon fraxineum.

The Ash-*leaved Tooth-ache Tree.*

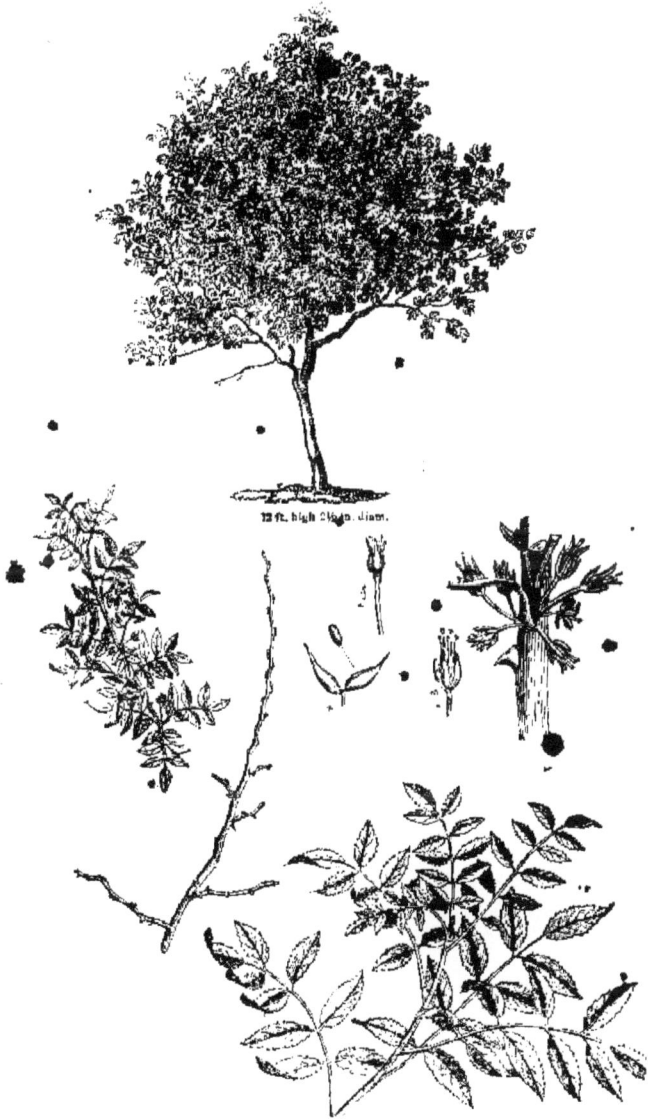

12 ft. high Clagh. diam.

Lightning Source UK Ltd.
Milton Keynes UK
UKHW020633260421
382641UK00004B/346